# The Globe's Upside &

# Trilocation -

# The 3rd Orbit of Earth

BY

KWABENA AMEN

iUniverse, Inc.
New York   Bloomington

## The Globe's Upside
## & 3rd Orbit of Earth - Trilocation

*iUniverse books may be ordered through booksellers or by contacting:*

*iUniverse*
*1663 Liberty Drive*
*Bloomington, IN 47403*
*www.iuniverse.com*
*1-800-Authors (1-800-288-4677)*

*ISBN: 978-0-595-52948-3 (pbk)*
*ISBN: 978-0-595-63001-1 (ebk)*

*Printed in the United States of America*

*iUniverse rev. date: 4/6/2009*

# Acknowledgements

My warmest appreciation to those who carved loving streams of knowledge in my mind: Supreme Almighty Intelligence, Leroy & Marjorie Searles, Dr G. Marcel, Wayne Williams, Benjamin Aird, Sammy Braithwaite, Dr J. De Vere Pitt, Desmond La Touché, Dr A. Friday, Justine, Nelson, Sonia, Yvonne, Hannah, Rawle, Melvin, Juchinar, Coslee and Auszamen. My gratitude goes out to many writers, thinkers, teachers, friends and families for their awesome ideas, synthesized here. *Nuff Love & Respect*!

This Book is dedicated to Earth's Children – *limit desires*

# Contents

# Figures

# Introduction

Images of Earth from space are fascinating. Perhaps, the third orbit of Earth, herein called *Trilocation*, and the up-side of the planet would not have been discerned without these remote perspectives. Notwithstanding, some ancient civilizations appeared to have had the knowledge that south was the top of Earth! How do we now know that south is actually upward? Why did we accept north as forward?

Seeking answers to this unquestioned assumption, initiated my exploration of the compass points, landscapes, climatology, and over two decades unfurled the concept of Earth's third orbit. I know now, that earnest, deep and persistent inquiry into any matter will yield valuable ideas or discoveries, once corresponding Spiritual depth is likewise enquired into.

The globe, as is displayed in most geography classrooms, is an all time admirable thinking tool, facilitating holistic views of Earth's structures on a scale that renders Earth's 6.5 billion people invisible. This large-scaled reduction of details enables hidden macro-patterns to be seen. These patterns are so extensive, that it took hundreds of years for Western cartographers to redraw the map of the world accurately. Even so, there remain inaccuracies in representations of the globe. One is the presentation of north as top of the planet.

The difference in perspectives between looking at a map of a country and looking at that country with a view from space is like looking at a mountain under-foot, and looking at the same mountain from the distance of another mountain. In the former, micro details of part is seen, while in the latter macro details of the whole is viewed.

In addition to the classroom, many media and business enterprise actively employ the symbol of Earth to establish scope, frame views and give the impression of universality. Marketing messages sometimes suggest that on the globe we are one human family, one harmonious planet, one market! What if these panoramic views are upside-down?

In the past, much human effort worked to ignore, escape and misrepresent certain Laws and Cycles of Earth. And while these have brought certain regions short term gains, they lead ultimately to long term stresses for the whole world. Over the last six centuries, people have developed lifestyles in disregard of, and in contravention to, many geographical and biological principles. For example, the light bulb has allowed us to break the laws of rest and sleep that was naturally regulated by the sun. We now have twenty-four-seven cities that no longer rest – a condition unseen in nature. When we holographically expand the consequences of unnatural lifestyles of modern cities into the foreseeable future, we end up with a dead Planet.

The cumulative effect of breaking Earth's Laws by so many people, now threatens our very survival. To behold the 22$^{nd}$ Century, 21$^{st}$ Century "civilization" must realign itself to Earth's cycles - find ways to work harmoniously with the planet by changing current imbalanced lifestyles. We must find new worldviews of ourselves that enable us to modify our destructive social and environmental footprints into sustainable patterns of living. We must acquire the correct perspective of Earth and all life forms on, above and within it.

Many aspects of our knowledge of life are limited - frequently inaccurate. For example, mankind does not behave as if he/ she understand that Earth is an alive being; as if he/ she comprehends that Earth is a *life form* that not only maintains its own life, but also sustains other diverse life forms upon itself!

One of the factors fuelling inaccuracy and imprisoning our understanding of life is our very definition of living beings. For instance, one condition of our current definition is that living beings reproduce themselves. Therefore, because Earth does not *seem* to sexually or asexually reproduce itself, it is not alive. But two concepts can help open this gate we have imposed upon our understanding. In the first place, let's suppose it takes Earth ninety billion years to reproduce itself, and let's assume that this reproduction is accompanied by the complete remake of all life forms currently existing on it, how then would *man* know that this reproduction has taken place?

Secondly, who holds authority to define life, is it the creatures, or their creator(s)? In other words, who authors the definition of artificial intelligence, is it the robot itself, or is it the creator of the robot?

There is but one *Life!* Everything taken in through our ten senses is alive! Earth is a living being! Planets, nebulae, air, water, etc., are all alive! Everything on Earth is therefore alive. When we raise our consciousness, we see that there are many *levels* of life, and many *forms* of life at various levels. There are manifested forms of life which range from gross expressions like bodies, objects, matter, etc., to subtle expressions like energy, electricity, gravity, etc.

In addition, there are hidden forms of life from which the manifested forms originate. These forms are hidden because they are not perceivable by the five outer senses. A scientific tradition that does not recognize, much less use the five internal senses, would have little information about these hidden forms. A stone is regarded by current science as a non-living thing. But we know that the activities within a stone at the sub-nucleic level reveals electrons and protons in rapid movement, exchanging electrons; bonding, and upholding relationships, recreating patterns according to characteristics that sustains the phenomenon of the stone! Furthermore, what of the intelligence that has organized the matter from gas into stars into stone? Moreover, imagine removing all (non-living) stones from the planet; what life would we have left?

Living and non-living classification of forms of life, as is currently practiced, is too limited a framework from which to truly answer the greatest questions of all times: *What is life? Who are we? Why are we alive?*

Psychologically distinguishing ourselves away from our connections with nature allows us to objectify Earth and its environment, and treat animals, plants, minerals, elements, etc., without care - *as if they were not alive.*

The mechanics of Generation, Organization and Dissolution of forms, surround us in contiguous patterns, habitats and cycles that illustrate the evolutionary progression of intelligence, within and among the various manifested kingdoms. From the simplest element, to moss, to the amoeba, to reptiles, to mammals, to man, we can comprehend a connection of energies within the apparent diversity of forms and habitats. Mankind depend directly and indirectly on other forms of life for survival. All forms of life depend directly on another form of life for survival. But there is order. Elephants do not depend directly on whales for their survival, nor do rabbits depend directly upon cats. What forms of life were mankind designed to depend on for survival? What is our order?

If mankind is the highest evolutionary form of life on the planet, and if life is infinite, then the evolutionary progression of people continues on another plane. This plane to which only humans have conscious access is called the spiritual plane. The mental realm includes the facility called *intellect* - our ability to choose actions, reflect on recorded data, react to and reorganize some of the laws and principles of the grosser expressions of life. We can choose to focus our intellect on the physical, or the Spiritual realm. The features of our "civilization" reflect the realm we choose to focus on.

In addition to analysing the grosser expressions of life with the external senses, we should also be discovering scientific knowledge of the subtler expressions of life, using the internal senses. This leads to social balance, morality, and into the realm of Divinity.

Current "scientific" knowledge is restrained. Fragmented definitions derived from employing only five senses to understand evidence, shuts down human understanding in some fields, even while progress is made in other fields. For example, governments spend millions to find cures for diseases such as cancer, in order to save people's lives, and that same government would spend billions on weapons that waste people's lives.

We must utilize the five Spiritual senses with our five physical senses, to properly understand life and re-align our perspectives correctly. Of what use will be our advanced technology and material progress, if greed and unrest destroys us all? Developing Spiritual senses lead to the reduction of greed, equitable redistribution of work, services and resources, and a foundation for peace among all nations.

Where science ends, Spirituality begins. After we analyze and divide nature in to the minutest unit possible, we come to a point where, for all practical purposes, our division of matter ends – this is the first stage of Spirituality!

The principles that are applied in understanding the science of Physics, for instance, are also applied in understanding the science of Spirit. The former analyses the physical realm, the latter analyses the Spiritual realm. Just as there are formulas, systems and features in the physical world, there are formulas, systems and features in the Spiritual world. Where correct conclusions are arrived at in any realm, progress can be made. The morality of decisions, choices and actions in physical life is determined by the decisions, choices and actions in Spiritual life.

Disconnections between science and Spirituality are modern mistakes. They are like disconnections between a tree and its reflection in a still pond. If we studied the reflection and ignore the existence of the tree, would we not be limited in our conclusions about what we see? At the edge of the pond the trunk ends and reflection begins. The trunk and its reflection are also connected through light - albeit *different* expressions of the *same* phenomena.

Let us take another example, *time*. There are four periods of time: past, present, future and eternal. We use most of our scientific energies to focus on: past, future and present, which are temporal, and ignore the eternal which is crucial!

Notwithstanding academic debate, a quality of contemporary science is its willingness to adjust to "new" insights and revelations.

A scientifically, industrially correct approach is for humans to live and work with the planet as a *Living Being*. Is it not better to redefine our definition of "life" and treat Earth with respect as a living being (Mother), rather than to maintain current definitions, and destroy ourselves in educated ignorance? Earth, like any other life form, has needs, limitations and conditions for her functioning and healthy existence. She has to interact with her natural environment. She has to coordinate sustaining forces in the ether; balance temperatures, gaseous, liquid and solid elements, etc. She has to provide food, habitats and resources for the life forms depending on her. Ignorant of these, humans extract, produce, consume and dump on Earth as if there are neither limits nor consequences; behaving as if they can exist independent of other life forms and ecological systems provided by Earth - living an upside-down life!

"Life" cannot come from "dead" things. Death is still taboo in current western culture. But, let's briefly reflect on another aspect of death. What dies? Is it Spirit or body? Since Spirit is eternal, it does not die! So it must be the body that dies, right? But how can we define a body as "dead", simply because it has been deserted by the primary Spiritual entity that occupied it? Isn't decomposition an activity performed by other forms of life – bacteria, worms, etc., – that emanates from within the body, recycling the gross matter of that body? Isn't it the same Spirit motivating the bacteria and worm to decompose, as had motivated the body? What then happens to worms and bacteria after their decomposition job is finished?

Earth is a massive generator of life forms. Earth's relationship with the elements, moon, sun, planets, stars, etc., has to be maintained within certain boundaries in order that life forms continue to exist on it, and humans have an active responsibility in this process, especially now that their activities reach across the whole planet.

Human forms are the greatest expression of life on Earth. The supremacy of human forms comes from the ability to reflect, react, respond to and transcend sensory stimuli. Unlike most other life forms, we affect the balance of life on the planet.

Ignorant of the huge responsibility of keeping the balance of life expressions on Earth (Garden of Eden metaphysics), mankind raped Earth in the name of comfort; at the expense of other life forms, including himself. Now the consequences of his actions have reached to the high physical heavens, and the planet is about to react. Extreme climatic and environmental changes globally, are nature's reactions to man's plunder. These reactions will force people to either redefine their philosophy and change current wasteful lifestyles, or perish in their obstinate mindset.

The impact of human activities on the globe is affecting it in ways previously unimagined. We have become incapacitated in our responsibility to sustain the planet with its various life forms intact, whilst harnessing from it, our needs. The reason for this is because of a vacuum in our practical Spiritual knowledge. After thousands of years on the planet, humans still do not recognize that the plundering, acquisition, development, false luxury, etc., we create, are useless to us in our moment of greatest need – our time of leaving Earth!

That mankind has an upside-down view of life is not just a literary anomaly; we have imposed a literal upside-down view of the globe upon our consciousness - perhaps the biggest geographical error since thinking that Earth was flat!

The summary effects of all human activities including: extraction, concentration, relocation and disposal of resources, in wasteful competition, that are harming Earth's being would soon lead it to re-balance itself - like a dog that scratches mites off its back! All life forms try to preserve themselves when threatened. Animal forms fight, hide or run away; plant forms withdraw or try to re-grow. Earth preserving itself would be interpreted by current science as "catastrophic, natural disasters". One example of Earth defending itself would be the churning of landmasses due to the consequences of mining, oil drilling, etc., as hundreds of miles of cavities created within Earth crust, in conjunction with crustal zones of weakness, collapse under tectonic movement, releasing massive quantities of heat and geostatic pressure into upper layers of the crust - the sort of conditions that sank Atlantis.

Some might argue that this is not a life form exercising its defence. But, let us listen to their learned arguments when their cities lie sixty feet under water, or sixty feet under fire; when a remnant hand-full of humanity would have survived an unimaginable disaster and would have learnt to live in growing isolation as their technology also died; scattered in pockets across the globe; until the time came that when they meet each other again, they would have forgotten that each other existed. They would call themselves' "strangers" and would have developed new sounds and letters in their languages; alternative ways of worshipping; different songs; various customs! They would have long forgotten, or have only mythological stories, of that day in the 21$^{st}$ Century AD, when Earth shook herself because the cancerous itch of human pollution had touched a nerve!

We must remember that the capacity of man to destroy himself and Earth is less than Earth's capacity to destroy herself and man! Only through respect of life in all its forms and energies, would we sustain ourselves, fulfil our responsibilities with beneficial consequences (intended and non-intended) to all relationships affected by our thoughts, words and actions; to *love all, serve all,* until we come in the consciousness of That Source that drives and defines us in this studio of time and experience.

Thus, if we continue to consume and waste resources as we did for the past recent centuries, we would provoke a reaction from Earth that would cascade a small percentage of humans, into the sticks and stones age, to restart civilization, *again!*

Fortunately, we can change universally to avert global catastrophe, or we can change personally to increase our chances of surviving the coming deluge. We can change our destructive footprints into life sustaining impressions, if we want to. We can practice correct perspectives of our planet and of each other; appreciate the wonder and frailty of Earth's universe; realize unity in diversity; understand the connectedness between *Relativity* and *Reality.* Of course, we have to start somewhere. Correcting our perspectives and representations of the globe and each other, is as good a place as any.

Earth's solar environment moves in a series of orbits around the galactic centre that are much alike that of electrons around a nucleus. Wind speed and directions, ocean currents, climate, etc., all have distinct patterns that reflect the operation of orbital forces on them. So too, the shape, movement and positions of dry landmasses have peculiar patterns. These patterns show another orbital force, *Trilocation,* as significant as *Rotation* and *Revolution,* acting to influence climate, continental shape, tectonic movement, atmospheric pressure, etc., across Earth's surface.

There is clear geophysical evidence revealing the operation of this third orbit, whose effects help sculpt the face of the planet and determine many natural patterns. This orbit renders the South Pole as the *front,* and the North Pole as the *back* of the planet.

In this book, representing over twenty years of theoretical geographical and astronomical research, I will try to present evidence of Earth's third orbit via features on its surface, explain the reasons for continental shapes, and account for certain characteristics and conditions on and around the planet, including reasons why south is upward.

It is hoped that readers would sense the underlying connection of life herein suggested, (like various coloured light bulbs, connected through electricity) and that this insight be used to help change current upside-down perspectives of the world - bringing us into better harmony with our geophysical mother, and with each other.

# 1. Incorrect World Perspective

Currently, north is regarded as forward or top of the world. This incorrect orientation of the globe has mis-informed social and scientific visualizations for centuries. In addition to the Earth's first and second orbits, *Rotation* and *Revolution*, its third orbit, *Trilocation*, determines that the south is indeed forward or top of Earth. How can we know this?

The Laws of motion and the direction of movement of bodies have resulted in conventions about perspectives that are applied uniformly in space and time, in order that we may understand our orientation and see where we are going. We clearly know which is the front of a bicycle, car, train, boat, aircraft, etc., and can distinguish their forward from their reverse movements. The front is the end that faces forward; that bears the brunt of airwaves, particles, pressure, etc., as the vehicle penetrates space. The rear end displays distinct characteristics of drag, temperature, pressure, turbulence and wave features distinguishable from the front. In the making of aircrafts, for example, the front and rear edges of the wings are designed to create differences in airflow that result in lift.[1]

In vehicles that are designed to move forward and backward in any direction like trains, the forces acting on the front and rear of these vehicles change depending on its direction. Designs of these vehicles differ to those designed to move mainly in one direction. What has all of this to do with South being the top of Earth? We will see that Earth shows which end is front from certain designs, features and conditions on its surface.

One region of Earth displays characteristics as those of the front of a moving body, whilst the opposite region displays characteristics as those of the back of a moving body. Patterns of landmasses and electromagnetic features likewise indicate sides.

We live in a world that contains many worlds, and is itself, contained by many other worlds. The microcosm world contracts infinitely into nothingness and the macrocosm world expands infinitely into nothingness. [2] Somewhere along this spectrum physical existence of rock, soil, plant, flesh, water, fire, air, etc., manifest and function for a period of time, then disappear out of the physical realm. This physical world consists of reflections, reactions and resounds of another realm which appears hidden to the five outer senses. As we better understand the micro, we better understand the macro, and vice versa. As we better understand the inner or hidden realm, we better understand the outer or manifested realm, and vice versa. Indeed, a grand holographic universe. To understand the inner realm we must use our internal senses.

# 2. Reasons for North as Up

There are no scientific bases for north being established as up or top of the World. In astronomical models of the solar system, for instance, planets are not arranged randomly according to the whims of the scientist constructing the model. They are structured with the Sun at the centre, followed outwards by Mercury, Venus, Earth, Mars, etc., according to their scientific order, revealed through research. [3] However, no such scientific principles appeared to have been applied (or known) when cartographers decided to present maps of Earth with north as up. It appears that ethnocentric, not scientific, reasons exist for our current perspective of maps with north as up. [4]

It is true that most of the dry land masses of Earth is in the northern hemisphere, since most people live here it would seem convenient to represent the northern hemisphere at the top of the globe for ease of viewing. But is this scientific? What about the true orientation of the planet? Ease, comfort of viewing, the fact that people living in the northern hemisphere printed most modern maps, should not be the reasons that north is located at the top, especially when there is scientific evidence to the contrary.

The cartographic "rule of ethnocentricity," - the placement of one's own territory at the center of a map - is an almost-universal feature of cartographic devices, including cosmic diagrams of pre-Columbian North American Indians; ancient Babylonia, Greece, China, and the medieval maps of the Islamic world and Christian Europe. This may be excusable for maps representing a small area or nation in the 12th Century, but when it comes to universal representations of the globe today, there need to be better scientific responsibility.

The practice of placing north at the top of maps came around the 15th century when European navigators started using the North Star and the magnetic compass. Maps were printed in Europe; therefore, the north became the top. Before this, East was oriented as up, perhaps due to the apparent rising of the sun and moon at that point. The Chinese (Orientals) however, put south at the top of their maps. Ancient Egyptians also referred to South as up. For example, Upper Egypt lies to the south while Lower Egypt lies to the north. [5]

Today, countries like Australia, Indonesia, etc., print maps of the world with south at the top, for reasons associated with tourism. These maps are referred to as *upside-down maps* of the world. However, this book offers scientific reasons why south is up, relative to a centrifugal and gravitational force that is orbiting the planet southward. South is in fact the globe's up-side.

Among other features, the third orbit of the planet explains why most of the land of Earth is situated in the northern hemisphere, and assists in understanding the shapes of Earth's landmasses, wind patterns, weather systems, as well as the magnetic skewing of compass needles northwards.

# 3. Trilocation – Premise

The Earth has a forward-facing side. It is Antarctica – the South Pole! How can we know this? The front of a moving body or object is determined by the direction of its movement. The features at the front of a moving body or object include cooler temperatures, higher pressure, and an outward from centre movement of wind waves that produces most turbulence at the rear. This forward-facing side is called the front.

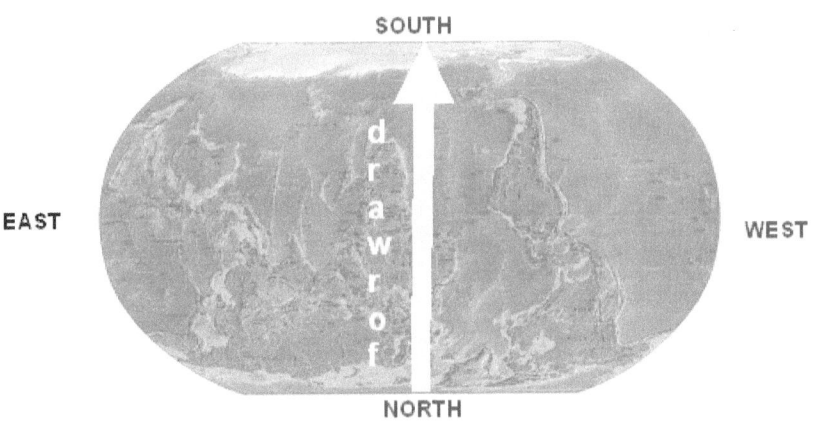

**Fig. 1 Trilocation Premise**

You would not get into a car, turn the ignition, shift into first gear, proceed to drive, looking out of the rear windscreen - orienting yourself as if the rear-end of your car was forward, would you? Yet, this is like what we do everyday when we display maps of the Earth with North as *up*. We look at the back of Earth as the front, and look at Earth's face as its tail. This limits our perspectives!

Here is evidence that Earth is on a third journey through space that renders Antarctica at the South Pole as up and forward of the planet.

The materials presented here are intensely condensed; it is hoped that the references given would provide readers with clarification and expansion of the principles and examples mentioned.

# 4. Forces Acting on Bodies in Motion

Forces that act on moving bodies, whether in the macrocosm or microcosm, mould features and influence conditions on the surface of those bodies. These moulding forces include: propulsion, gravity, friction, drag, resistance and inertia. These forces, activated by Trilocation on a global scale, shape the landscape, climatologic and other features on Earth. It is the characteristics of these features that indicate the direction of this third orbit of Earth.

Gravity is force acting on Earth that keeps all objects stationed on the planet; it gives them weight and affects shape during growth or emergence. [6] Earth is influenced by several sources of gravity. Some of these are Earth's own magnetic field, and gravity from the moon and sun. A free, round object will roll downhill because of gravity. Rivers, waterfalls, rain, sleet and snow would not occur without it. Life forms owe their sculpture and strength partly to gravity.

Matter moves according to waves; the lighter, or less dense the matter the easier it is to discern the motion-patterns they display. The wave patterns of smoke, fire, water, mudslides, lava, molten metal, etc., display similar flowing characteristics. The shape of these matter-in-motion is not only determined by the force that initiated the movement, but also the forces of inertia, resistance, friction, and drag that affect the mass as they move.

Inertia is the tendency of a body to preserve its state of rest or uniform motion unless acted upon by an external force. Resistance is the opposition to flow of matter. Friction is a resistance encountered when one body moves relative to another body with which it is in contact. Drag is the resistance to the motion of a body passing through semi-solid, fluid, or air.

As an object travels forward, these forces act within, on and against the surface of the object. Collectively they shape the object over time. It is therefore possible to decipher the direction of movement of matter from the characteristics and pattern of change that its shape displays even after it has stopped moving, or even if its movement is extremely slow.

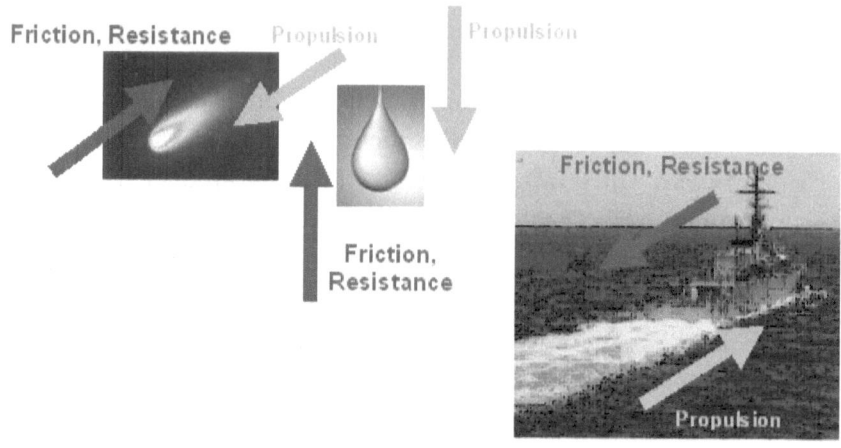

**Fig. 2 Forces Acting on Bodies**

# 5 Shape of Flowing Matter

The shapes of all flowing matter like fire, water, lava or molten metal, etc., have distinct bell-shape characteristics that are influenced by the forces acting on them relative to the direction of movement. Let's make a closer examination of the movement of matter before we look at certain structures on the planet. For example, when muddy water droplets splashes onto a car while it is moving, they are blown backwards along the car's body, in bell-shape patterns - according to the effect that inertia, drag, etc., have on them. Other semi-solid matter, flowing down a gradient, show distinctive and consistent bell-shape patterns, for example, lava flows or mud slides. Such bell-shape patterns are reflected in the shapes of the planet's major landmasses. The bell-shaped patterns of continents, with compacted bulges towards the north and elongated tails to the south, illustrate a downward flow of landmasses from south to north.

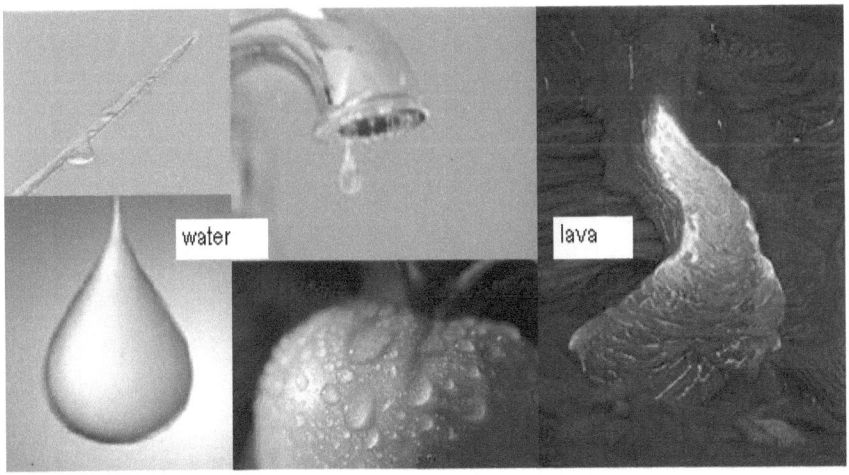

**Fig. 3 Shape of Flowing Matter**

When landmasses were initially belched out from within the body of the planet, during its period of early cooling, some 4.6 billon years ago, matter eventually congealed to form continents and islands. During this young cooling era, "continents" would have "floated" on the relatively hotter and more liquid mantle, with greater velocity than by the time Earth cooled enough to allow plant life to appear.[7] The direction in which continental shelves with their cargo of dry land floated was from south to north - as evidenced by projections of the movement of Pangaea (original collection of continents) into becoming the five continents, islands and tectonic plates, etc., on Earth's surface.[8] See figure 4 below. Why did landmasses on the surface of the earth travel from south to north, and still continue to do so? What is motivating this mass movement?

# 6. Early Shape of the Continents & Plate Tectonics

Most of the landmasses on Earth's surface were initially nestled at the South Pole. However, over millions of years, they separated and travelled northwards due to the effects of Trilocation upon the tectonic plates.

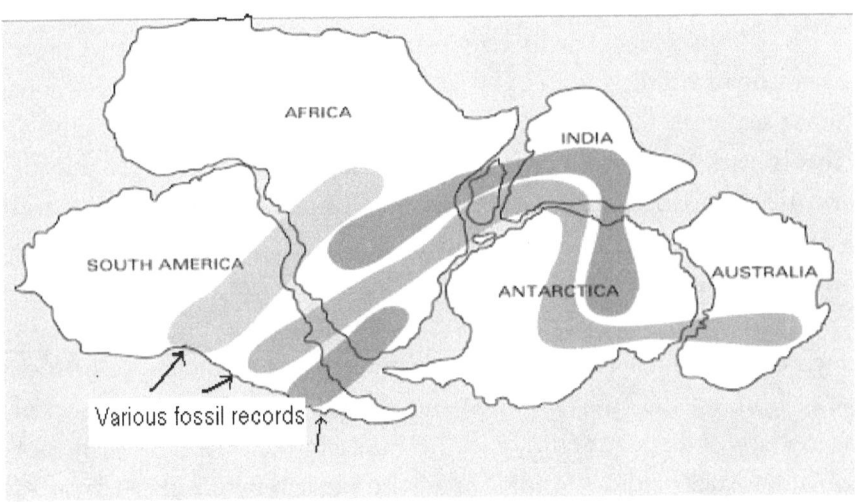

**Fig. 4 Early Shape of Continents**

On the map below, (figure 5) it is seen that the drift of continents is from the southern hemisphere into the northern hemisphere. [9] The direction of this drift is due to gravitational drag on the continents as a result of the southward thrust of the planet. Trilocation is a southward movement with a corresponding northward drag on the continents and their tectonic plates, just like the forward movement of a car would cause mud splashed onto its sides to be dragged backwards.

As Earth plunges through space with the south in front, inertia, friction, drag, together with the convectional forces within the mantle and core of the planet, create a backward push on the landmass on the surface of the planet. It is more than likely that the convectional currents within the planet are also skewed northwards by Trilocation. This force is similar to the inertia one feels when an aeroplane is taking-off and one's body is pressed back into the seat. It is also similar to the way in which the force of wind would push back your hand, if your stick it out of the window of a fast moving car.

Now magnify this principle to a 25,000 miles circumference vehicle, moving through space at 250 km per second. The constant pressure of similar forces pushes the mobile continental masses back towards the north. [10] The northern hemisphere can be regarded as the collection zone of Earth's landmasses.

It is known that Earth's crust varies in thickness, and that it more or less floats on the mantle. [11] The greater the landmass that sticks out above sea level, that greater the thickness of the crust at these points. This is because a greater degree of buoyancy between the crust and the mantle is needed under mountains than under flat or low lying land - rather like ice floating on water; the larger the piece of ice above the surface, the deeper the piece of ice under the surface.

There are cycles of emergence, circulation and submergence of rock and crust material on Earth's surface. Earlier emergence of rock mass from Earth's interior in previous cycles includes the landmasses of Greenland, North America, and the Eurasian continent.[12] Emergence of "new" rocks from the interior of the earth now occurs in areas of volcanic activity. For example, Mt Pelee in Martinique, Caribbean, which in 1902 spewed out enough molten rock and ash to cover the city of St Pierre. [13] New rocks can also be seen being formed in Kilauea volcano, Hawaii. [14] Since the earth is much cooler than it was four billion years ago, and since the crust is much harder and solidified, we do not see emergence of new rock on a scale that produces continents. (Indeed if the earth was still producing new rocks on that scale, it would most likely be uninhabitable for animal forms of life.) It is mainly in the joints between tectonic plates that we see most volcanic and earthquake activity today.

The oldest rocks on the surface of the earth are found in Canada, Greenland and Northern Europe, i.e. the landmasses surrounding the North Pole – furthest away from the South Pole. [15] This is because after emerging from the interior of the earth in the southern hemisphere, they travelled northwards over millions of years, due to the effect of Trilocation. As more rock material later emerged from the region of the South Pole, they age would be relatively younger.

Earth's crust is composed of a series of plates that move relative to each other across the planet's surface. [16] Joints occur between and among plates and certain features occur at these joints depending on the relative direction of their movements. The material of the earth's crust can be regarded as *floating* upon the material of the earth's mantle.

It has been known to western science since the 1950's that continents move. However, only in the 1980's has it been recognized that smaller pieces of continents can move independently, in some cases for thousands of kilometers, between and over major continental joints. These far-traveled pieces of the edges of continents are called *exotic terranes*. They range in size from small cities to small countries. Pieces of Alaska, for example, have probably come from as far away as Chile. [17]

Figure 6 shows the major joints or boundaries of continental plates. In some areas, for example between the African and American continents, the plates are moving away from each other, or extending. In other areas, like between the pacific and Australian plates they are moving towards each other, or compressing, and still in other areas like between the Pacific and North American plates, they are sliding along side each other.

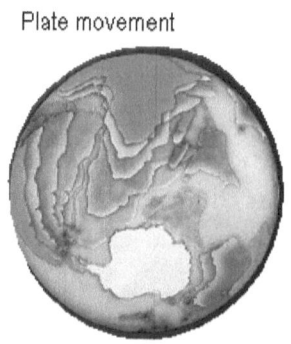

Plate movement

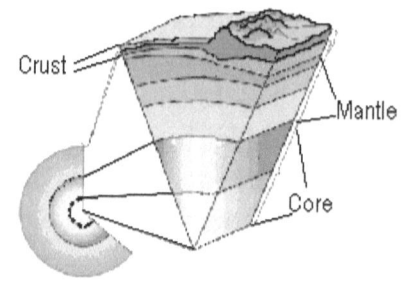

**Fig. 5 Structure of Earth**

Geologists have attributed the movement of these plates to the movement of the magma in the core; [18] to some extent this is so. But what is initiating these movements in the magma and outer core? It is suggested that initiation originates from the Trilocation of Earth that produces south to north drag, and the Rotation of Earth, which initiates west to east centrifugal shifting. Trilocation's southward plunge creates a drag on the crust through friction, which pulls major pieces of tectonic plates backward, in the same way as passengers are pushed backward when a vehicle accelerates. This drag significantly affects currents within the mantle and core which act beneath the crustal plates.

At the time when landmass materials were being churned out from the molten interior of the Earth, the emerging, solidifying rock mass took their south to north bell-shapes partly from the effect of Trilocation; partly from the resistance and friction on the masses as they flowed northward. What we witness today are bulges in continental material at the head of the flow and elongation at the rear (bell-shapes). This is the reason that the northern-most tips of continents are wider than their southern-most tips. This pullback of resistance has sculpted the plates into their present shapes, which are consistent with the shapes of most flowing matter. See figure 10.

# 7. Structure of Earth

The basic structure of Earth's body, with several zones of rock and magma in varying degree of viscosity, temperature and composition allows for continuous movement of continents, land masses, rocks and minerals across, into and out of the interior via tectonic joints, geostatic currents and pressures. [19]

While the pressurised inner core of Earth is hot and molten, the outer surface is generally cool and solid.

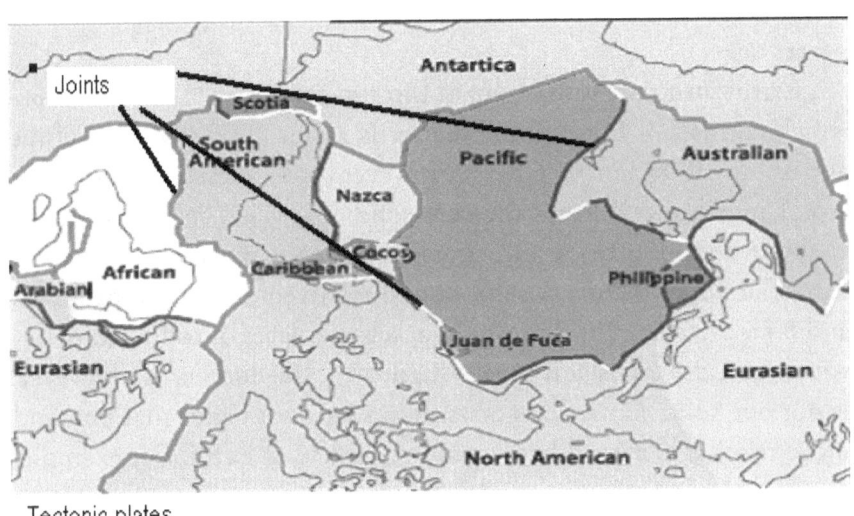

Tectonic plates

**Fig. 6 Plate Tectonics Boundaries**

Figure 4 shows one pattern of what Earth's crust may have looked like when the land masses emerged from the early, cooling era of the planet. Compared to its rotational speed and surface temperatures a few geological moments after birth, Earth has slowed and cooled considerably. [20] This slowing and cooling of the planet separated planetary materials – elements, minerals, compounds, etc., - according to their density, and allowed the sculpture of dry landmasses according to certain geo-physical forces that acted on them. The patterns sculpted on the continents, with their bell-shape traits, assist us in deciphering Earth's third orbit. Let's take a closer, comparative look.

A moving drop of water carrying dust particles, that is moving down a gradient will have at its "bottom" these entrained dust particles. This is because they are denser than water. They will "emerge" at the bottom-most part of the drop. Likewise, compared to the mantle, the denser rock and soil material that formed the dry land of Earth's crust first emerged from the interior at the South Pole, because of the gravitational propulsion of Trilocation southwards. Most likely, it was here that the relatively heavier continental masses were "belched" out to later become dry land, whilst the planet was in a relatively molten state. As more landmasses matter were belched out, they travelled northwards, and continued to do so after solidifying, although at a slower rate.

Earth's structure is made up of three major zones. [21] The *inner core* which consists of hot, molten matter is about 5000 degrees Celsius and about 3000 km in radius. However, great pressure presses the atoms together and keeps the core solid. The *middle zone* or *mantle* is about 2880 km thick and consist of viscous magma in a state of flux. The upper mantle is solid due to relatively cooler temperatures and high pressure. The *crust* is the outermost layer of solidified rocks, which host dry lands and oceans (figure 5). The outer crust, however, is not one solid continuous mass, but a series of plates that join and move relative to each other across the surface of Earth. For example, the San Andreas Fault line in North America, where the North American and Pacific plates are sliding adjacent to each other in a south-north fashion. [22]

The crust is made up of solid material but this material has not the same distribution everywhere. The oceanic crust average about 9 km thick and mainly consists of heavy basalt rocks, and the continental crust averages about 30 km thick and is made up of mainly lighter granite rocks. [23]

The nature of these three major zones allows convection currents, geostatic pressures and the overriding force of Trilocation to shift the outer crust-plates over the mantle. The drift of continents on the surface of the planet, mainly from south to north, is therefore a phenomenon occurring because of Earth's third orbit.

# 8. Movements of Objects in Space

All objects in existence are moving relative to each other, around orbital centres, and within systems of bodies with shared orbital centres. For example, on a macro level, Earth is firstly rotating on its axis at 0.5 km per second. Secondly, Earth is revolving around the Sun at 30 km per second. Thirdly, Earth and the Solar System are moving around the Milky Way Galaxy at 250 km per second, and the Milky Way is moving at approximately 630 km per second. [24]

On a micro level, a nucleus moves within a blood-cell, [25] that cell moves within the blood plasma, and the blood stream travels around the body, even as the body moves about on the ground in search for food, shelter, etc.

Thus, bodies are involved in many cycles and orbits, either spinning on their own axis or revolving around certain centres, repelling or attracting others, most of which are above or below the sphere of awareness. These cycles and orbits influence the lives of everything in existence; in fact they are integral to life itself. It may be impossible to find a stationary object in space; in fact, space itself may also be moving. Working in tandem with astronomical orbits and cycles produce harmony.

# 9. Rotation of Earth

The Rotation of Earth about its axis, tilted 23.5 degrees perpendicular to the Sun, is once every twenty-four hours. [26] Since the circumference of Earth at the equator is 25,000 miles, Earth rotates once in about 24 hours, or just over 1000 miles per hour. This speed is less at the poles for the same reason that a compact disc rotates slower at its centre than at its edges. The effects it produces includes day and night, and with the moon's gravity, tides.

The moon is rotating on its own axis and revolving around Earth. Four types of gravitational forces inclusively operate on Earth: the gravity of Earth on itself; the gravity of the Moon on Earth; the gravity of the Sun on Earth; and the gravity of the Galactic centre (along the trajectory of Trilocation) on Earth.

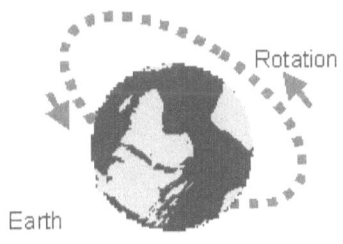

Fig.7 Rotation of Earth

Due to the centrifugal force of Earth's Rotation, there is a "bulge" at the equator, established when Earth began its rotational movement. It is suggested that the angle of Rotation at 23.5 degrees is in line with the trajectory of Trilocation. Trilocation's trajectory through space is of course like a spiral. This spiral is also known as the wobble of Earth in space.

# 10. Revolution of Earth

Earth revolves around the Sun as its solar centre, once in every 365.25 days, or a year. [27] The most noticeable effects of Revolution are the changing seasons and annual cycles. Earth is moving around the Sun at about 67,000 miles per hour. The varying durations of sunlight which are most pronounced in Polar latitudes are the combined effect of Revolution and the angle of Earth's Rotation.

**Fig. 8 Revolution of Earth**

The Revolution of Earth is responsible for the seasonal variation that affects climate, erosion, weathering, plant growth and animal migrations. Revolution also affects human behaviour and consumption of energy. In the winter in the northern hemisphere, for example, more resources are required for heating, visibility, mobility, etc., in order that people produce in similar quantities as in summer. In nature, hibernation of plants and animals in synchronicity with winter, allow them to reduce activities and conserve energy, thereby sustaining their habitats and lifestyles in the longest term. Mankind, with so called superior intelligence, shows little regards for these cycles of seasons and have invented gadgets and medicines that allow continuous, and in many cases increase, commercial and industrial activities in winter. The accumulated effect of this battle against the Revolutionary cycle is stressed-out people as well as planet. The claim of superiority of intelligence is not reflected in the collective human treatment of their environment, or each other.

If humans were to work with the cycles of nature more closely, for example, reduce activities to sunrise and sundown during winter, they would be more in-tuned with the cycles of Earth and have more harmony in essential activities. Some of the mental and economic fears that maintain current out-of-cycle human societies are examined in Chapter 24.

# 11. Trilocation of Earth

With Earth, the Solar System, is circling the galactic centre in a direction that renders what we call the south of Earth as forward.

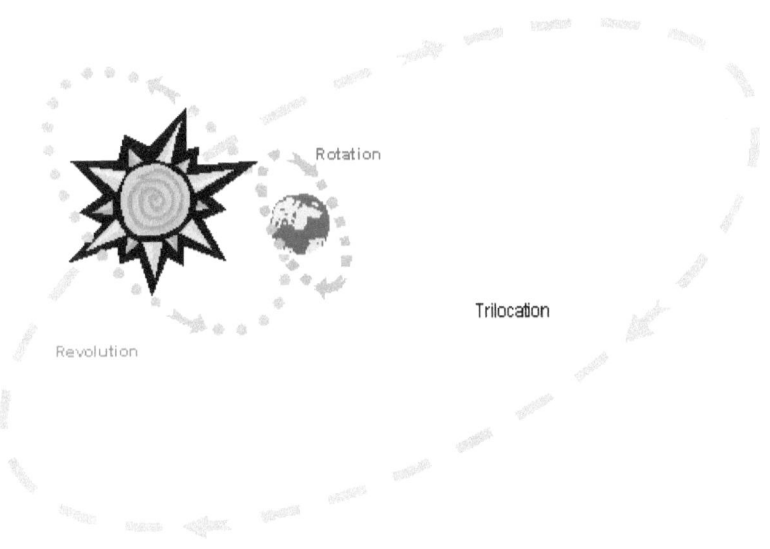

**Fig. 9 Trilocation of Earth**

Furthermore, the galaxy is circling within a group of galaxies through time and space along a fourth orbit, and so on. Trilocation indicates that since the nine planets and their moons remain in relatively constant proximity to each other, the trajectory of the solar system within the galaxy is in a direction that renders the South Pole of Earth as forward. Several pieces of evidence, when taken together, show this third orbit.

# 12. Shape of Major Land Masses

Earth's dry landmasses are not randomly shaped. Tectonic plates, continents and other major landmasses display patterns that indicate a "flow" from southern to northern hemispheres. The shapes of major landmasses are reflected in the shapes of flowing matter, as discussed earlier. The path of Trilocation, as suggested, is consistent with the shapes of continents.

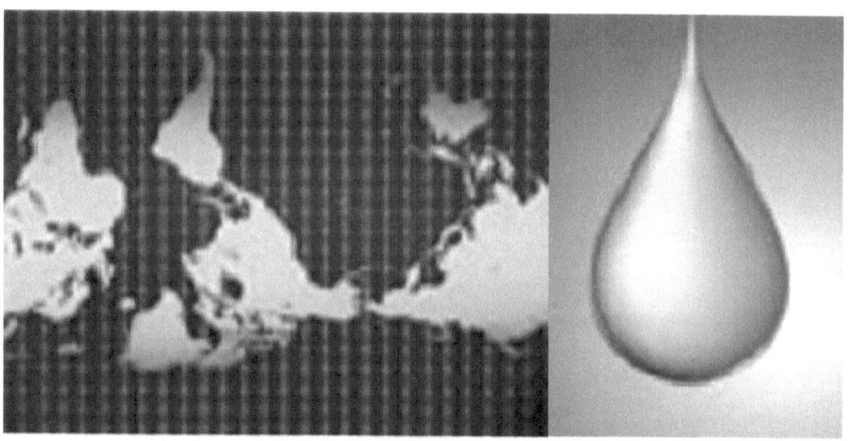

**Fig. 10 Shape of Major Landmasses**

The bell-shaped characteristic of flowing matter is replicated on Earths crust. If we look at the pattern of all major landmasses (and tectonic plates), we would clearly see this bell-shape - as if the continents were "flowing" northwards.

# 13. Accumulation of Landmasses in Northern Hemisphere

Trilocation gives reasons for the phenomena of the accumulation of landmasses in the northern hemisphere, compared to the southern hemisphere. If the landmasses of the crust were floating on the mantle (and core) of Earth, and if Trilocation operated to push landmasses backward, relative to its trajectory, then over millions of years we would expect an accumulation of landmasses in one of Earth's hemispheres. Exactly this phenomenon has occurred in the northern hemisphere, which is the receiving half of Earth's mobile crust.[28] As the globe plunges through space, south-forwards, the resulting drag on the crust pushes continents northwards resulting in an accumulation of landmasses there.

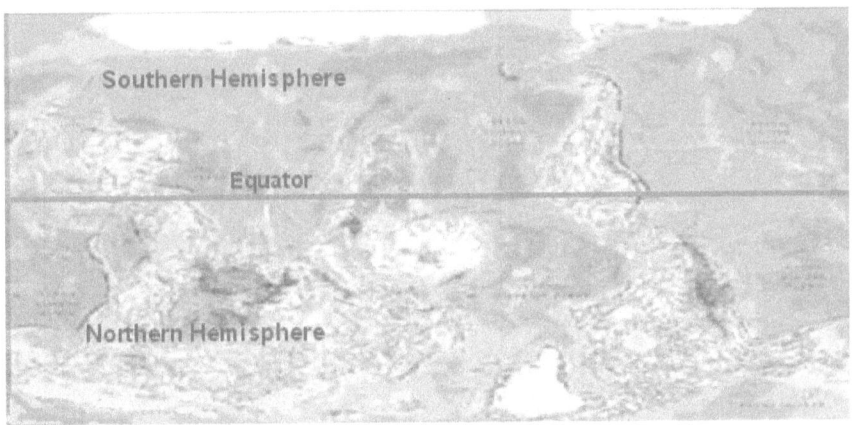

**Fig. 11 Accumulation of Landmasses in the Northern Hemisphere**

# 14. Cooler South Pole

As Earth travels South-forwards through space, blasts of cold temperature currents from outer space cool the atmosphere over the South Pole more than any other part of the globe. Temperatures at the South Pole can be minus 89 degrees Celsius while temperatures at the North Pole are minus 32 degrees Celsius. [29] These cooler conditions at the South Pole are produced because of the movement of Earth through space with South as forward.

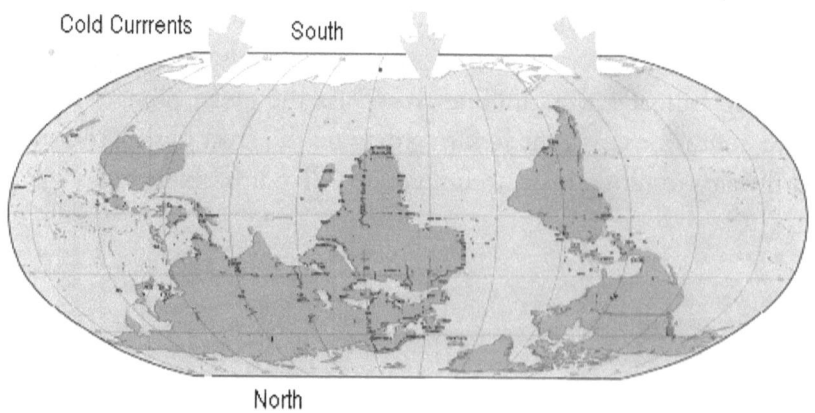

**Fig. 12 Cooler South Pole**

The back of a moving object is relatively warmer than its front. The same is true of Earth where the front (south) is twice as cold as the back (north).

# 15. Larger Accumulation of Ice at South Pole

Since cold galactic and solar temperatures are constantly cooling the South Pole more than any other part of Earth, much larger ice mass has accumulated there over millions of years. The ice at the South Pole is more than twice the area of ice in the North Pole. Antarctica's ice mass is thicker, deeper, windier, and has a smaller zone of melt during summer seasons relative to the Arctic (figure 13). There are grassland and animal life in the Artic region that are not apparent at similar latitudes in the Antarctic. [30] Indeed the Artic is more of a frozen ocean that largely thaws in its summer, while the Antarctic is solid ice (and landmass) throughout the year. It is easy to perceive that extreme cold conditions are encouraged at the south because of the Trilocation movement of the Earth through space with south at the front.

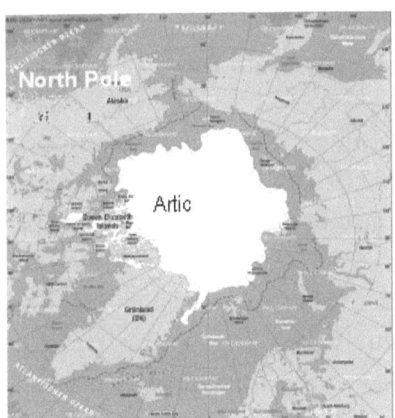 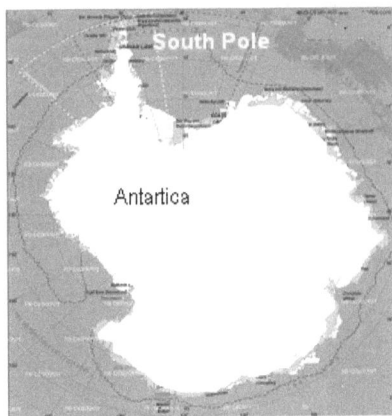

**Fig. 13 Larger Accumulation of Ice at the South Pole**

The south is being cooled far more than the north (the rear). In addition to the poles being angled away from the sun, the cooling effect of solar and galactic temperature currents at the South Pole, give rise to its unique frozen characteristics.

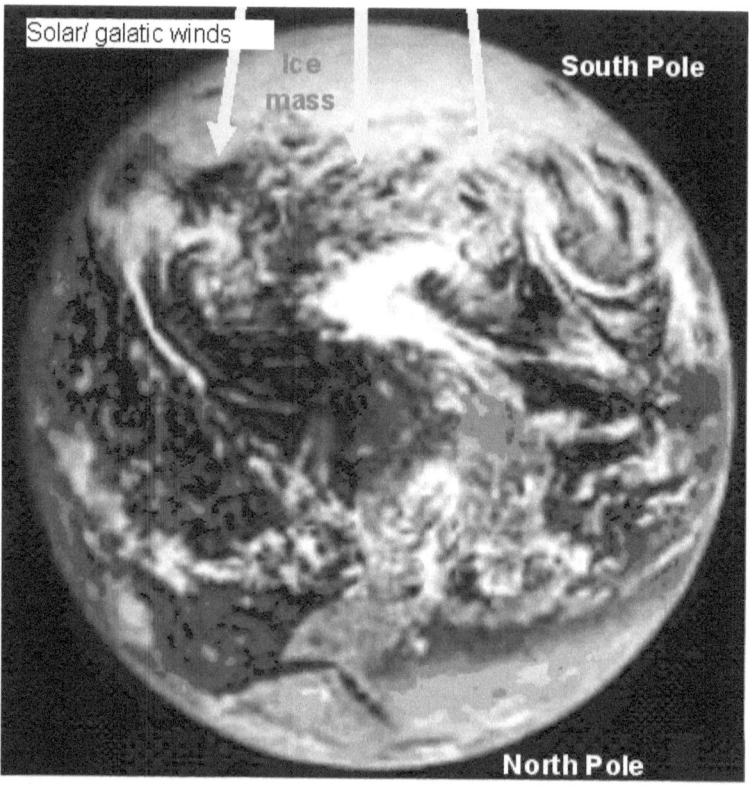

**Fig. 14 Satellite View of Ice at South Pole**

This satellite view of Earth (figure 14) shows the greater accumulation of ice at the South Pole - the front of Earth. This frontal exposure results in Antarctica being the thickest ice continent, the windiest, driest and most desolate place on Earth.

When riding a bicycle, you feel the cool wind on your face, chest and fingers much more that on your back. Any vehicle travelling through the air will be relatively cooler at the front than at the back. Earth's journey through space in Trilocation's direction is no different. Cold solar and galactic currents (temperature zones) hit the South Pole more than any other part of the globe, making Antarctica the windiest and coldest place on Earth.

Polar Regions are generally cooler than Temperate or Equatorial Regions due to the quantity of sunshine received. However, the South Pole is doubly cooler than the North Pole as a result of it plummeting forwards through space. It is the front of Earth on its third trajectory. Antarctica is therefore its driest and iciest frontier.

Another way of looking at the planet, in relation to the cold and heat that it receives from space and the sun respectively, is to imagine it as a spinning ball at an intersection point on a grid: with a south-north movement of cold, and an east-west movement of heat.

# 16. More Pronounced Winds in the Northern Hemisphere

Wind currents and drag pressures build-up in the northern hemisphere, and fluctuate relatively greater than in the southern hemisphere due to Trilocation's effect on Earth's atmosphere. Intense winds and oceans currents reflect the south to north drag. These exists and fluctuate in annual cycles, linked to the seasonal "movement" of the Sun between the Tropic of Cancer and Tropic of Capricorn, located 23.5 degrees north and south of the Equator, respectively [31]

**Fig. 15 More Pronounced Winds in the Northern Hemisphere**

Jet Streams are wind currents, which twirl around our planet in great circles, roughly around similar latitudes. These giant, *river of winds* can be several hundred miles wide, 1 to 2 miles in depth, can be found between

12,000 to 80,000 feet above Earth's surface, and can reach speeds of 400 miles per hour. [32] There are five major Jet Streams, but their distribution is more in Earth's receiving northern hemisphere – due to the third orbit. Polar Jet Streams occur in both hemispheres; Sub-Tropical Jet Streams occur in both hemispheres also. However, the Equatorial Jet Stream only occurs in the northern hemisphere. See figure 15.

Warm, rising wind currents from the equator, travel to higher northern latitudes than they do towards southern latitudes. The reason is that atmospheric drag from the south front forces wind systems backwards - towards the north.

It is suggested that the location of the Equatorial Jet Stream in the northern hemisphere is due to drag on Earth's atmosphere generated by Trilocation. This drag effect also explains why the Polar Jet Stream in the southern hemisphere (average location 30 degrees south), travels in an arc wider than their counterparts in the northern hemisphere (average location 45 degrees north). See figure 16.

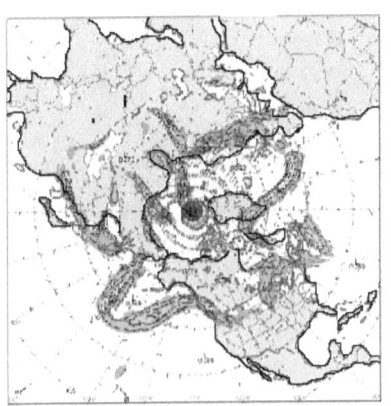
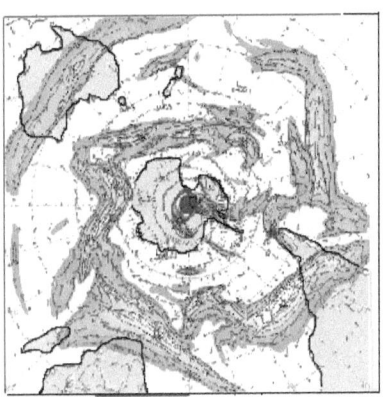

Polar jet streams in Northern hemisphere

Polar jet streams in Southern hemisphere

**Fig. 16 Polar Jet Streams (in September)**

# 17. Paths of Severe Whether Systems

The general paths of hurricanes, typhoons, tornadoes, and monsoons - severe whether systems - which occur in reaction to heat and pressure changes in the atmosphere of the planet, are from east to west, and south to north. [33]

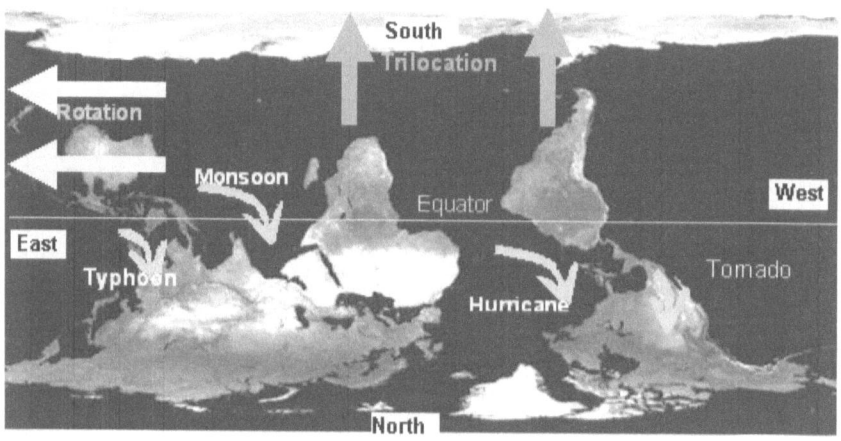

**Fig. 17 Paths of Severe Weather Systems**

Their paths can be generally looked at as moving diagonally from southeast to northwest. If there were no common forces influencing the direction of these severe weather systems, we would expect them to move randomly over their regions, or according to local winds patterns, but this is not the case. A force, greater than the Northeast and Southeast Trade Winds, keeps these severe weather systems along predictable paths. What is witnessed is that not only do hurricanes repeat the general movement (westwards and northwards) of each other, but so do all other severe weather systems.

It is suggested that the effect of Rotation skew the general direction of all severe weather systems westwards, while the effect of Trilocation forces them northwards.

Why are severe weather systems located usually just north of the equator between latitudes 0 and 30 degrees? It is suggested that the eddying effect of Trilocation "winds" as they pass the south-north curvature of the planet at the equatorial circle, in conjunction with the north-south apparent movement of the Sun and the Inter Tropical Convergence Zone in September, is what produces these eddies - turbulent, spinning, electromagnetic characteristics - of intense weather systems. The spiraling winds trap heat and water and condense them under low pressures; eventually exploding and losing their moisture, heat and electricity, until at last they fade as they travel into cooler northern latitudes, where the intense turbulence of equatorial region is abated. Of course, hurricanes, etc., have their annual cycles. If the ozone layer's protective role is reduced (due to pollution), then it can be expected that increased galactic exposure of the south of the planet would influence more intense and frequent storm systems in the northern equatorial regions.

# 18. Shape of Earth's Atmosphere

Observe the shape of a passing comet, or a drop of water moving down a slope, or the wake of a ship sailing on the sea. You see similar characteristics in the shapes produced which is generally ball-shape at the front with tail-shape at the end – determined by the direction of movement. Earth's atmosphere is similarly shaped due to its south-first movement. The atmosphere in the northern hemisphere is 'thicker' than the atmosphere in the Southern hemisphere. See figure 17. What features on Earth show this?

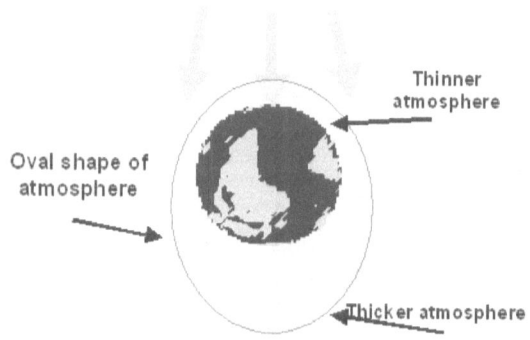

**Fig. 18 Shape of Earth's Atmosphere**

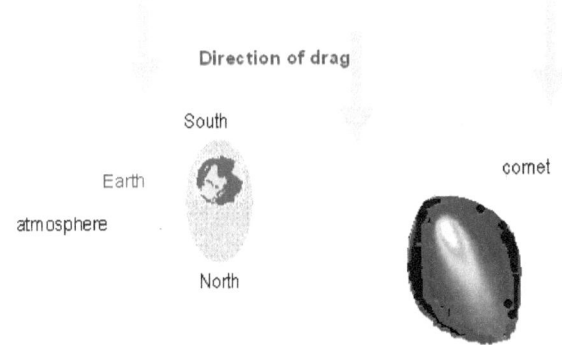

**Fig. 19. Shape of Comets**

Let us take a look at what happens as Solar and Galactic electromagnetic "winds" rush against the planet at hundreds of kilometers per second. In the case of a comet, we see that the resistance of space at its front creates a shorter distance between its front edge and its core, than the distance between its tail and its core. [34] The frontal pressure exerted against the (gaseous) atmosphere of the comet forces it into an oval or bell-shape. Similarly, the atmosphere on Earth at the South Pole would be skewed - thinner in the south and thicker in the north, if the earth was plummeting through space southwards.

This frontal pressure causes the ozone layer (atmospheric layer) at the South Pole to be relatively thinner than the ozone layer at the North Pole.

# 19. Holes over Antarctic

Because of the thinner atmospheric layer over the South Pole, in a situation of depleting levels of ozone, "holes" in the ozone layer would appear first over this region, before any other part of the atmosphere. [35] For example, imagine a car tyre with grip that is unevenly layered, having a thick layer of rubber on one side and a thin layer on the other side. As the tyre is worn down, the part with the thinner layer of rubber would show the underlying threads first.

Again, we can see the shape of the ozone layer reflected in the shape of a comet. Since the comet's tail extends further away from its center, than its head, the layer of gasses at its head would be thinner compared to the layer of gasses at its tail. This shape is due to resistance against the comet as it travels forward through space. These same principles apply to Earth's atmosphere.

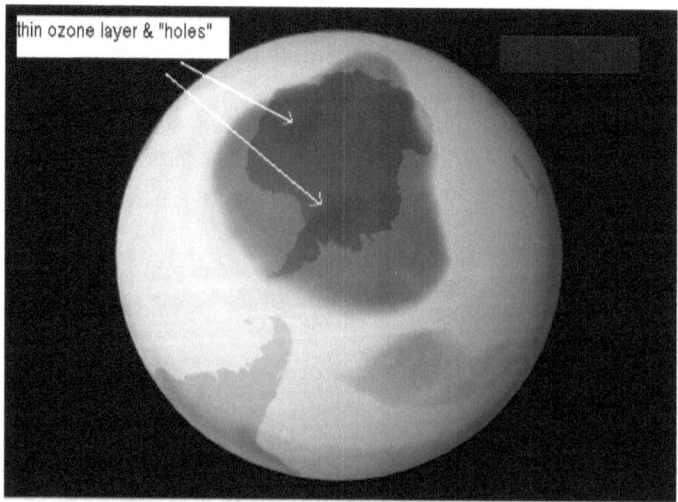

**Fig. 20 Thin Ozone Layer & "Holes"**

Satellite thermal images illustrate the relatively thin atmosphere over Antarctica; figure 19. Here, forces of friction and resistance impact the atmosphere over the South Pole. If the ozone layer is egg-shaped, and the thinner part is over the South Pole, then "holes" would appear there first. Thinning of the ozone layer due to pollution, reduces its cushioning effect and allows increased light, heat and pressure to influence the polar ice-cap (Antarctica) thereby accelerating melt. Of course this takes place in conjunction with the greenhouse effect.

# 20. Northern & Southern Lights

The impact of friction and resistance against Earth's upper atmosphere produces, together with the magnetic charges at the poles, a series of electro-static waves and particles that refracts the light of the Sun at polar latitudes in a manner that produces the astonishing phenomena of the Auroras. Like the molten matter moving within the earth's mantle and core, the atmosphere also produces its own electrical and magnetic energies. [36] The Auroras dance across the evening sky in a manner similar to the turbulence in the wake of a passing ship. The dancing lights reflect electromagnetic activity concentrated at polar latitudes.

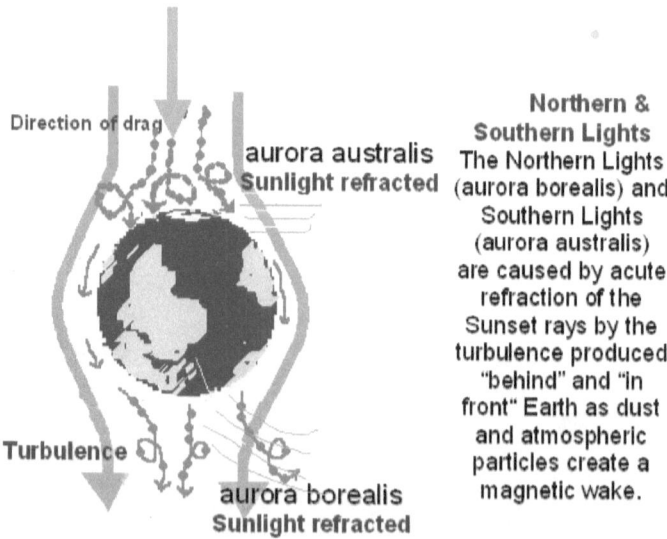

Direction of drag

aurora australis
Sunlight refracted

Turbulence

aurora borealis
Sunlight refracted

Northern & Southern Lights
The Northern Lights (aurora borealis) and Southern Lights (aurora australis) are caused by acute refraction of the Sunset rays by the turbulence produced "behind" and "in front" Earth as dust and atmospheric particles create a magnetic wake.

**Fig. 21 Northern & Southern Lights**

**Fig. 22 Displays of Aurora Borealis**

Figure 21 shows images of the Aurora Borealis (Northern Lights) in the northern hemisphere, usually visible in polar latitudes. That the Auroras occur at polar latitudes is explained as being due to the turbulence in the electrostatic atmosphere as a result of Trilocation. Just as there is more turbulence at the front and mainly the rear of a vehicle in a wind tunnel, there is more turbulence at the front and rear of the planet – the south and north polar regions respectively. It is suggested that the electromagnetic eddying of particles in the polar atmospheres, is what produces the effect of the Auroras.

The dazzling display of auroras are produced in the electro-static wakes of the planet, as much as the wake of a ship produces white foam in the turbulence at the stern and the bow.

# 21. Northward Pointing of Magnetic Needles

The consistent swinging of magnetized needles towards north is fascinating. A degree of mystery and magic accompany our first-time witnessing of this "hidden" force operating on the planet.

The magnetic compass reliably guides navigation because magnetic north can always be determined. It is one of the oldest instruments used in navigation and has been a vital tool on land, air and sea. [37] Compasses allow ships to remain on selected courses. By taking bearings of visible objects with a compass, the navigator is able to fix a ship's position on a chart. Of course, GPS technologies now precisely locate positions on the globe for navigators.

The origin of the compass lies among our ancient un-recovered past. Certainly, the Egyptians knew about the properties of magnetism over 4,500 years ago. The four corners of the Great Pyramid of Giza are almost perfect right angles and align to the four points of the compass. [38] Some 2,000 years ago, the Chinese were aware that an iron bar stroked with a lodestone acquired a north-south directional property. By the 10th Century AD, probably via traders from China or Egypt, the idea appeared in Italy.

Although the magnetic compass was in general use since the Middle Ages, science knew little about precisely *how* it always pointed north. Now, it is understood that a compass works like this because the planet is like a giant magnet generator, [39] with a huge magnetic field between its two poles. If this was all there is to the principle, we would expect that a negatively magnetized needle would point south and a positively magnetized needle to point north, or vice versa. But this is not the case. Whatever the polarity, magnetized needles continue to point north.

Magnetized needles tendency to always point north is due to drag against the Earth that trains the flow of particles in the electromagnetic fields backwards to the north. As Earth plummets south-first through space, its magnetic fields are skewed northwards, and correspondingly, the consistent north-pointing of all magnetized needles.

**Fig 22.1 The Magnetic Compass**

# 22. Other Evidence of Trilocation

With awareness of the third orbit of Earth, many geological and astronomical phenomena can be better understood, including but not limited to the following geophysical characteristics, which are outlined in brief:

a. That the texture of the landmass at the Northern ends of continents display wide, crushed, sunken coasts and sunken valleys, while the textures of landmass at the southern ends of continents show raised smooth coasts and relatively pointed shapes. [40]

b. That the perpendicular angle of the Earth's rotation at 23.5 degrees to the sun is the angle of Trilocation. [41]

c. That landmasses generally have their highest mountains located north, in front tectonic boundaries. [42]

d. That the general paths and direction of major cloud systems is from south to north, followed by intense reverse reaction. Jet streams are more intense in the relatively pressurised northern hemisphere. [43]

e. That the highest tides at Nova Scotia, in the northern hemisphere, are due to pressure build-up of ocean currents from the south. [44]

f. That stargazing is better in the southern hemisphere than in the northern hemisphere because there is less atmospheric electrons, dust, turbulence, etc. These particles bounce-off Earth's surface and are blown backwards to the north. [45]

g.  That meteoric showers are generally more pronounced in northern hemisphere in northern winter, when warm rising atmospheric particles from the southern hemisphere are blown backward (north) creating a denser atmosphere (than when it is winter in the south) that provides increased collision opportunities for transversing meteorites and cosmic dust.[46]

h.  That the Acasta Gneiss in the Canadian Shield in the Northwest Territories, Canada, contains Earth's oldest rocks because they have been exposed to the troposphere the longest, having travelled the most across the surface of Earth, on their south to north journey. [47]

i.  Trilocation predicts that flights into outer space would be more fuel efficient and incident proof, if departures from Earth's atmosphere took place from the northern-hemisphere, and re-entry into Earth's atmosphere took place in the southern-hemisphere.

# 23. Trilocation - Action & Limitation

The study done for this book is largely theoretical, and is yet to be confirmed mathematically and corresponded astronomically. If further research shows the discovery of the third orbit of Earth to be misinformed, then the evidence synthesized here nonetheless reveal a hidden phenomenon operating on Earth, that has influenced and continues to influence it in ways suggested, that warrants further research.

Future editions of this book will include results of new investigations, mathematical, geological and astronomical formulae, expansion or findings in support of, or in challenge to the existence of Earth's third orbit - Trilocation.

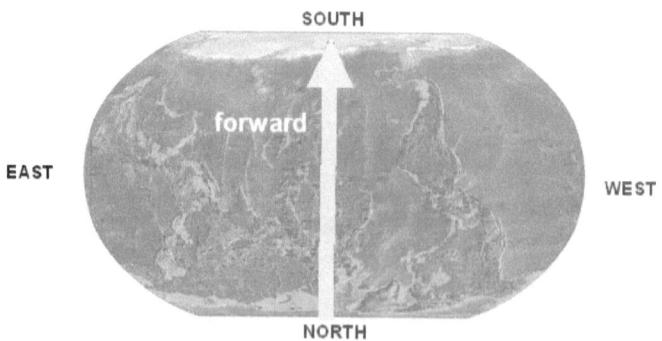

**Fig. 23 Trilocation Action**

## Global Environmental Destruction indicates that something is wrong with human perspectives

Humans are changing the composition of the atmosphere in a manner that has no known natural parallel.

Pollution-related closings and health advisories at U.S. beaches were more numerous than ever in 2005, according to NRDC's annual report. Across the country, there were more than 20,000 days of closings - an increase of 5 percent from 2004. NRDC expects the upward trend to continue.

Pollution is a major problem world wide and appears to be on the increase. Reports of blue-green algae outbreaks, ever increasing soil salinity, exhaust fume levels, chemical farming practices, mining practices and beach pollution are occurring more often. These pollutants have a harmful effect on all living things - plant, animal and human.

### Global Warming - the worst in history

**Fig. 24 Global Environmental Destruction**

## The Upside Down World!

Despite the colossal advances in science and technology, human beings are unable to find peace, balance and harmony. Investment in warfare lead to mass destruction. Investment in right conduct will sustain life.

**Fig. 25 The Up-Side Down World**

# 24. Conclusion - Correct World Perspective

Correct the World! Maps, globes and representations of the planet should be displayed with south at the top, not because of social accidents or ethnocentric ideas, but because scientific evidence shows this to be fact. Our incorrect perspective of the World has contributed to skewing precepts of humanity and of life that has ultimately resulted in intolerable levels of corruption, inequity, poverty, crime, failures, insecurity, war and economic meltdown on a global scale, at the same time that ecological systems of the planet is seriously threatened. Yet our scientific, political, religious, economic and educational institutions fail to see the connection. They do not see that the solution to this physical disharmony is Spiritual - the *practice* of peace, love, care, truth, full employment, free access, equality, unity and universality.

How does our incorrect perspective of Earth contribute to our upside down practices in Society? To put it another way, how does misunderstanding about the planet affect our relationships with each other, other forms of life, and matter, inclusively?

Let us circumnavigate the subject of Trilocation for a moment, and examine one of our unquestioned assumptions. Let us question the practice of urban distribution of *essential* goods and services. What if food, shelter, medicinal care, education and career opportunities were provided freely, or made available to all through a method other than finance? And what if this philosophy led to a great peace and social harmony among nations and neighborhoods, in ways previously unimagined, would not our current practice of buying and selling necessities be an upside-down practice of correct distribution of essentials within civilized cities?

The knowable universe is vast.

Matter on Earth is limited.

The mind can conceive desires that the fixed resources of Mother Earth cannot produce nor satisfy. For example, imagine how far the solid and liquid matter comprising the earth's physical structure would extend into space, if they were gathered and packaged into one-meter cubes, which were then stacked one on top the other. Now, imagine someone attempting to execute this task – mobilizing people, resources and machinery for the purpose!

Yet, this is the kind of economic vision being practiced in global society in relation to the world's non-renewable resources, where deprivation of necessities for some, is see-sawed by gluttony and wastage of luxuries by others; where deposits of minerals are extracted, concentrated and the by-products dumped as poisons; where species and habitats are destroyed, as if there are no consequences; no limits. The application of incorrect perspectives about life caused this. We practice inequitable economic principles, based on legal greed and socially accepted wrongs, ignorant of the eventual destructive effect on the whole. The poor do not understand their poverty; the rich do not understand their inequity. Yet, the cure for this malady can be found in the ancient injunction to *"limit desires"*, and the golden rule to *"love each other as self."* But popular marketing messages preach just the upside-down - *"more, better, faster"*, and *"treat yourself exclusively"*, in direct opposition to principles of sustainability and holistic care.

What then are the benefits to human society of realigning global perspectives? Facilitating the breaking of mental chains that would allow the reawakening of another level of consciousness among leaders and decision makers, by providing a platform for a paradigm shift in the way they understand and organize their societies.

There is no doubt that we need to correct our application of science and technology in relation to life forms sharing the planet. The best relationship we can have with animals, for example, is to leave them and their habitats alone! Animals thrive best in their own homes, rather than in labs, cages, zoos, race courses, reserves or farms. Just because they speak different languages and organize their resources differently is no reason to abuse them. We must learn to live harmoniously with them. Furthermore, the design of the human body is for vegetarian fuel!

This consciousness comes with true knowledge of self and respect of life. We need to re-deploy personal and global resources from destructive to constructive mechanisms. Free guns, bullets and bombs must be replaced with free food, shelter, medicine and opportunity, in order that the fundamental purpose of human living be achieved.

Love.

99 % reducing of wastage, reusing packaging, recycling materials, limiting consumption and renewing resources will sustain environments for posterity. The cancerous messages in the mass media must be replaced with messages of satisfaction, balance and truth. If you cannot switch off all the lights of your cities at night and rest peacefully as a society, then you are not civilized, not free and certainly not at peace! Civilized societies are aware that every mother waits anxiously for her offspring to return home every day; every father wants to provide shelter for his family; every child likes to go to sleep loved and well fed; and such societies provide these basic civilities universally!

Instead we have crowded people into metropolises that emphasize differences rather than sameness, manufacture half-truths, and stimulate destructive fantasies. Relationships are polarized and based on identification of self with body, or if schooled, mind, when in fact, identification ought to be based on Spirit. We thus have an upside-down view of our own Selves. We are not matter, nor mind. These are the instruments and energies we use. We are Spirit in a human form. And we are the only forms in this solar system that can realize the magnificence of the One Omniscient, Omnipotent and Omnipresent Being, whose quality, but not quantity, we share! The difference between a human being and GOD is like the difference between a drop of sea water and the ocean.

We err when we identify our Selves with the tools through which we operate in the mental and physical planes. Identification with transience is mistake. Misrepresentation of life, due to misidentification of Self is the greatest mistake we ever make! An upside-down model. The Truth is we identify correctly with Eternity - the Spiritual plane!

Not knowing Truth identification, mankind group themselves under various organizations and labels, then work, trade and war with each other, according to differences they condition themselves to perceive between the needs of their bodies, minds or organizations, and the needs of the bodies, minds or organizations of *others*. They identify themselves with the articles of their institutions. And identify their opponents according to their labels. This is rather like a computer describing itself according to its installed programmes. There is nothing wrong in forming organizations. In fact, we could not survive without them. However, the policies within groups must give equitable benefits or beneficial consequences to everyone and everything that comes within their spheres of influence. In this way they become institutions meeting the standards of universal survival. This standard may seem difficult only because it is not much practiced. Wherever these practices exist, Oneness is experienced.

Modern societies act against their Oneness. What are *right* in society are the remnant principles of Oneness that cannot be suffocated through legislation, or social prejudice. No matter how inappropriate a piece of legislation may be, it is purported to be hinged to a *right* that the legislators are trying to achieve - no matter how un-grounded they may be about that right. History has shown that whenever the prevailing social climate is negative, right surfaces in order that peace and balance be restored. Wherever right is unable to re-surface, conflict continues.

Despite colossal advances in science and technology, humans are unable to find the peace, balance, harmony or happiness touted by behaviour shaping institutions. Sophisticated concepts of civil-lies and superiority complexes bury justice and right. Entrained investments in military and industrial complexes motivate economies and lead to mass destruction on many levels. The cultivation of barbarism within pillars of civilization has turned much good discoveries and progresses into instruments and strategies of pain, destruction, and violence in many spheres of human activity. Their impact upon elemental, mineral, vegetal, animal, and global expressions of life is wanton recklessness.

49

Through misconception and misrepresentation of life, human forms attempted to distinguish and separate themselves from everything alive and have put cages, cameras and microscopes onto manifested things searching in transience for transient answers to transient questions, when the knowledge they truly seek lies just half an inch within their chests, accessed by quiet moments of reflection, contemplation and meditation! But instead of this earnest self-inquiry, they spend billions, kill millions, waste resources; searching in the *effect*, for the *cause;* searching the physical for the Spiritual; walking down-river looking for the source!

What if south is indeed upward, would this not mean that current views of geographical models of the earth are upside-down? Have we not also been looking at society upside-down in relation to manufacturing instruments of war to secure peace, rather than manufacturing instruments of peace to secure peace? Modern societies would freely give a bullet that kill, but they charge for a plate of food that sustains life. And this is called security. From the perspective of the correct distribution of resources for civilized survival, this is as upside-down as it gets!

The first requisite to destroy something or someone is to disconnect them from oneself or one's interests. It is a mental and emotional process that has little to do with the reality of a situation and more to do with the concoctions of ego. Once something is regarded as distinct or separate, it can be treated as such. The more we know of our connections with someone or something the more we care and sustain that relationship. Any human male can mate with any human female and their offspring would be a human child. Everyone laugh and cry, moan and groan, in the same language. Therefore, there is only one human family, but ethnocentric barriers prevent humans from acting on this truth within their interdependent, international, socioeconomic activities. Instead, we witness class, race and creed wars wasting precious limited resources and hurting people.

The collective human world, at the beginning of the 21st Century is thus, conceptually upside-down on the physical level, practically upside-down in the social plane, and identified upside-down in the Spiritual realm. One mis-step trips the other steps. Of course there are pockets of right knowledge across the globe, but these are not yet the prevailing practices in modern cultures.

To further understand our connection with life, we must look at the point of unity of matter and energy. Everything subsists in Spirit as One. Therefore, no differentiations exist. Everything is experienced as One – Truth – Consciousness - Bliss - Eternal. Some cosmological traditions have labeled this state of Unity of Spirit as Atma, Amen, Nirvana, Enlightenment, etc. However, in order that experiences occur, Spirit must separate itself from itself - into energy and matter – to provide opposites and spectra. The energy/ matter duality continuously subdivides into many hidden and manifested forms and energies (life) that we experience/ insperience. Correspondingly, there is Spiritual spark within all matter and energy. *Individuality* in life denotes an undivided unity, acting in duality, in order to have *experience*. The scientific path of *insperience* leads to unity. [48]

One of the Laws of life discerned through synthesis of secular and Spiritual knowledge is that every manifested body or activity has a corresponding metaphysical body or activity. According to the ancient wisdom, *"...as above so below."*

To be able to treat correctly every other human being and life forms, etc., that are sharing the planet with us, we must understand the connection and unity of life. This unity does not exist in the realms of matter nor mind. It exists in the realm of Spirit, and is understood when one searches this inner realm, not superstitiously, but scientifically. Once we achieve this state of awareness, we behold the reason for distributing basic necessities to all. Many decision makers can intellectually reason out some principles of unity, but they find it impossible to put them into practice in the office or community. This is because a true Spiritual understanding has not occurred. A true Spiritual understanding leads to practice in one's life of the Truth experienced, and provides resolutions to obstacles.

We know that if we approach the study of an object or body with incorrect perspectives, we will reflect incorrect conclusions about that object or body - until we discover the mistake.

Modern societies have misapplied their knowledge and technological advances against the planet and against each other. Truly civilized, free, democratic and fair societies use their trade surpluses to provide the necessities of life to all, so that everyone can pursue their paths of growth, purpose, service, happiness, etc. *("...feed my sheep...").* Where would crime statistics be if everyone had free, responsible access to the resources they needed? Such societies also recognize and exercise their responsibilities to animals, vegetation and the environment.

What else defines freedom?

Public funds, charitable donations, taxes, profits, surpluses, etc., are to be used to provide and maintain necessities for all. Instead, we see finance diverted to fund projects that abuses nature, construct prisons, manufacture weapons and facilitate the massive mobilization of people and resources into researching wars and strategies of control of *others* through unjust national, socio-economic practices. Institutions of force expands because the models upon which their societies are based are upside-down! If our behaviour shaping institutions – homes, religions, schools, legislatures, etc., - were shaping free societies, through equitable, universal policies, there would be no need for weapons. Armies were originally created by the fathers of civilization as a source of arms (assistance) coming to us in our time of need – to build our homes, prepare our farms, execute our ideas, etc. Instead, governments now have thousands of young men and women trained to deliver sophisticated instruments of pain and death, the execution of which later silently destroys them and their relationships with post traumatic stress disorders, and governments claim that this is defense! Upside-down?

Surplus monies should be used to provide the basics, at the highest quality, free to all. Total change of human perspectives and practices are needed. Everything we have been told about ourselves and society needs questioning. Human society has the technical and logistic capacities to do right. It is decision-makers' vision that is lacking. They are busy attaching labels to others and treating them as such. There is a *way,* but there is no *will!* This lack of will is due to lack of unity of thoughts, words and deeds. Because of their upside-down identification with duality rather that Divinity, people are unable to find unity, peace, love.

Citizens are conditioned to view *other* nations, *other* people, *other* races, *other* languages with which they do business as *different*, and to a limited extent this duality is useful – provides humour and variety. But, when it comes to the re-distribution of necessary goods and services, and when it comes to finding universal grounds within our diversities, in order that we relate harmoniously, the primary consideration should be: *"Is this a human being or thing, connected with me to the experience of Life? If so, how can I help?"*

Business and trade should be done as among friends in attitudes of equality, cooperation, caring and sharing. We can create financial markets that cater for the prosperity of a few at the expense of the majority, or we can create financial systems that cater for the prosperity of all. Complimentary businesses are better than competitive businesses because everyone benefits. All of Earth's wealth has already been created. No matter how much money we have accumulated, whether through honest or dishonest means, we will not take one coin with us through our private gate of death. Therefore, the philosophy underlying our financial systems is correct when everyone everywhere prospers.

According to principles of economics, the balance of economies occurs when budgets record *zero* at the end of the financial accounting period. Yet, some countries suffer huge debts, with corresponding huge surpluses in others; which are then reinvested into more imbalances through vicious cycles of competition. Mankind does not behave as one sustainable, complementary, connected economic community where budgets are balanced among all trading partners. Why does the UN Economic and Social Council exist? If this seems like a simple economic view, it is! Truth is simple. Lies are complicated. Competition, manipulation, deceit, corruption, secrets and confrontation are sophisticated strategies employed by governments and institutions to create products and services upon the chaos resulting from their inequity. Complimentary trading, care, honesty, consideration, etc., in all human relationships, among all cultures and creeds, brings solutions; brings true peace.

We are Truth.

We are Divine Beings having human experiences.

However, our choices have been led astray through misidentification, misinterpretation and misapplication of many natural cycles and principles. For example, our greatest invention – the light bulb – has become our greatest darkness. Greed for synthetic light has empowered us to break natural cycles of sleep and rest and create twenty-four-seven, artificial lifestyles that produce stresses, wastes and diseases. Greed for synthetic light and the wonders of electrical wizardry is destroying Earth's ecology. For every one socio-economic problem solved, two others are created, and this is called progress. Upside-down principles?

Returning to the subject of Trilocation; we have seen that certain geological patterns carved on Earth's surface, reveal a third orbit of the planet and illustrate some of the effect of this trajectory on its environment as it travels southwards through time and space. This orbit determines that the South Pole is the top of the Earth. A change in our perspective of the planet may support a paradigm shift in respect of our approach to life. If we got the upside of the globe wrong, is it possible that we have also gotten other assumptions of life (and each other) wrong?

We can correct our perspective of the Earth! Human forms may then adjust choices, seek Truth, find balance, and reach the second stage of Spirituality - aligning identity with the Supreme Almighty Intelligence.

# References

1. *The Work of Wings*, http://virtualskies.arc.nasa.gov/aeronautics/tutorial/wings2.html

2. *Theories of Macrocosms and Microcosms in the History of Philosophy*, G. P. Conger, NY, 1922, which includes a survey of critical discussions up to 1922.

3. *Solar System*, http://en.wikipedia.org/wiki/Solar_system

4. *The Northern Hemisphere*, Prof. David Deley, http://members.cox.net/deleyd/religion/solarmyth/nh.html

5. *Discovering Ancient Egypt*, http://www.egyptologyonline.com/introduction.htm

6. *Gravitation*, http://en.wikipedia.org/wiki/Gravity#References

7. *An Overview of Earth History*, http://en.wikipedia.org/wiki/Gravity#References

8. *The History of Continental Drift - Alfred Wegener*, http://www.bbm.me.uk/portsdown/PH_061_History_b.htm; *The Changing Earth*, http://books.google.co.uk/books?id=zE5-d1IiqxEC&pg=RA1-PA641&lpg=RA1-PA641&dq=a)%09The+texture+of+the+landmass+at+the+Northern+ends&source=web&ots=06XeirzZgQ&sig=JH4W-57WZ8pdwGACaW4jawviLqQ&hl=en&sa=X&oi=book_result&resnum=1&ct=result; page 641.

9. *The Physical Environment - Tectonics and Landforms* http://www.uwsp.edu/geo/faculty/ritter/geog101/textbook/video/exotic_terrane_video.htm

10. *Landmass in Northern Hemisphere,* http://wiki.answers.com/Q/ Percentage_of_land_mass_in_northern_hemisphere

11. *The Earth's Interior,* http://www.solarviews.com/eng/earthint. htm; Beatty, J. K. and A. Chaikin, eds. *The New Solar System.* Massachusetts: Sky Publishing, 3rd Edition, 1990.

12. John Roach, *Oldest Rocks...,* http://news.nationalgeographic.com/ news/2008/09/080925-oldest-rocks.html

13. *Mt Pele Volcano,* John Seach http://www.volcanolive.com/pelee. html

14. *Hawaiian Volcano Observatory,* http://volcano.wr.usgs.gov/ kilaueastatus.php

15. John Roach, *Oldest Rocks...,* http://news.nationalgeographic.com/ news/2008/09/080925-oldest-rocks.html

16. *Introduction to Earth Movements and Structures,* http://www.uwgb. edu/dutchs/STRUCTGE/EarthMvts.HTM

17. *The Early Earth,* http://www.earthmuseum.segs.uwa.edu.au/__ data/page/48874/The_Rock_Cycle.doc

18. *The San Andreas Fault,* Dr David K. Lynch, http://geology.com/ articles/san-andreas-fault.shtml

19. *Earth's Structure,* http://mediatheek.thinkquest.nl/~ll125/en/struct.htm

20. *Milky Way - Velocity,* http://en.wikipedia.org/wiki/Milky_Way# Velocity

21. J. S. Giamportone & W.G.W. Booker, *How Plates Move,* http:// www.platetectonics.com/book/page_4.asp

22. *The Physical Environment - Tectonics and Landforms* http:// www.uwsp.edu/geo/faculty/ritter/geog101/textbook/video/exotic_ terrane_video.htm

23. J. S. Giamportone & W.G.W. Booker, *How Plates Move,* http:// www.platetectonics.com/book/page_4.asp

24. *Milky Way - Velocity,* http://en.wikipedia.org/wiki/Milky_Way# Velocity

25. *Cells*, http://www.elcamino.edu/faculty/kvillatoro/bio10_lectures/cellstructure2.pdf

26. *The Earth's Rotation*, http://www.windows.ucar.edu/tour/link=/the_universe/uts/earth2.html&edu=mid

27. *The Earth's Revolution*, http://www.boscobel.k12.wi.us/~schnrich/eath's_revolution.htm

28. *Mercator projection*, www.nationalatlas.gov/articles/mapping/a_projections.html

29. www.antarcticconnection.com/antarctic/weather/index.shtml

30. *Artic Wildlife*, http://www.saskschools.ca/~gregory/arctic/Awildlife.html

31. *Circle of Latitude*, http://en.wikipedia.org/wiki/Circle_of_latitude

32. Jose Reyes, *Influential Jet Streams*, http://cubanology.com/Articles/Influential_Jet_Stream.htm

33. http://www.bbc.co.uk/weather/features/understanding/hurricane_isabel_2003.shtml

34. *Comet Hyakutake*, http://astro.umsystem.edu/andy/astrop/96b2/96b2.htm

35. University of Cambridge, *Ozone Hole tour*, http://www.atm.ch.cam.ac.uk/tour/part2.html

36. http://www.exploratorium.edu/learning_studio/auroras/happen.html

37. *Magnetic Compass*, http://www.encyclopedia.com/doc/1O225-magneticcompass.html

38. *Measurements of the Great Pyramid*, http://www.repertorium.net/rostau/measures.html

39. *Planet Earth, a Great magnet*, http://www.spaceweathercenter.org/swop/Science_Briefs/Magnet/1.html

40. *Map projections*, www.nationalatlas.gov/articles/mapping/a_projections.html

41. *The Earth's Rotation*,

42. http://www.windows.ucar.edu/tour/link=/the_universe/uts/earth2.html&edu=mid

43. http://maps.google.co.uk/maps?hl=en&q=globe&um=1&ie=UTF-8&sa=N&tab=wl

44. Jose Reyes, *Influential Jet Streams*, http://cubanology.com/Articles/Influential_Jet_Stream.htm

45. *Some interesting Nova Scotia Tide Facts*, http://museum.gov.ns.ca/fossils/protect/tides.htm

46. *Southern-Hemisphere Stargazing*, http://quamut.com/quamut/stargazing/page/southernhemisphere_stargazing.html

47. *Major Meteor Showers*, http://meteorshowersonline.com/major_meteor_showers.html

48. *Oldest rock on Earth*, http://en.wikipedia.org/wiki/Oldest_rock#Oldest_rock_on_Earth .

49. *Man's Reality*, http://www.srisathyasai.org.in/Pages/His_teachings/Mans_Reality.htm

## Other References & Credits

- *Astrophysics Science Division (ASD) at NASA's Goddard Space Flight Center*, http://astrophysics.gsfc.nasa.gov/
- www.EnchantedLearning.com
- www.solarviews.com
- www.fotosearch.com
- http://imagine.gsfc.nasa.gov/index.html
- http://pubs.usgs.gov/gip/dynamic/continents.html
- TA Marryshow Community University, Tanteen, St George's, Grenada.

www.ingramcontent.com/pod-product-compliance
Lightning Source LLC
Chambersburg PA
CBHW021008180526
45163CB00005B/1930